Transport and Energy Processes 2022

Held at the 2022 AIChE Annual Meeting

Phoenix, Arizona, USA
13-18 November 2022

ISBN: 978-1-7138-7912-1

Printed from e-media with permission by:

Curran Associates, Inc.
57 Morehouse Lane
Red Hook, NY 12571

Some format issues inherent in the e-media version may also appear in this print version.

Printed with permission by Curran Associates, Inc. (2023)

AIChE
120 Wall Street, FL 23
New York, NY 10005-4020

Phone: (800) 242-4363
Fax: (203) 775-5177

www.aiche.org

Additional copies of this publication are available from:

Curran Associates, Inc.
57 Morehouse Lane
Red Hook, NY 12571 USA
Phone: 845-758-0400
Fax: 845-758-2633
Email: curran@proceedings.com
Web: www.proceedings.com

TABLE OF CONTENTS

ADVANCED ELECTROCHEMICAL ENERGY STORAGE TECHNOLOGIES I

CO2 CAPTURE, UTILIZATION, AND DISPOSAL: KEY TO CLEAN ENERGY PRODUCTION

POSTER SESSION: TRANSPORT AND ENERGY PROCESSES DIVISION

ADVANCED ELECTROCHEMICAL STORAGE

THERMAL ENERGY STORAGE

TRANSPORT IN ADVANCED FUEL CELL TECHNOLOGIES

ADVANCED ELECTROCHEMICAL ENERGY STORAGE TECHNOLOGIES II

531a Structure and Dynamics of Water-in-Salt Litfsi Electrolytes from First-Principles Molecular Dynamics Simulations

Ramanish Singh, Xiaobo Lin, Yong Zhang, Edward Maginn, Peter Cummings, Joern Siepmann

Concentrated aqueous solutions of lithium bis(trifluoromethanesulfonyl)imide (LiTFSI) have emerged as a safer alternative to the widely used organic-based electrolyte solutions for electrochemical energy storage in batteries. In the present work, we have performed first-principles molecular dynamics (FPMD) simulations to study the water network and ion solvation structures and their dynamics in aqueous LiTFSI at various water/salt ratios. The structure factor calculated from the simulations shows good agreement with the experimental results. Analysis of the ions' solvation environments reveals that the lithium cation predominantly interacts with four oxygen atoms from different water molecules and TFSI anions, forming a near-tetrahedral geometry with lithium at the center. At lower salt concentrations, this first solvation shell is preferentially populated by three water molecules and one TFSI anion, but changes to two water molecules and two TFSI anions at higher salt concentrations.

Hydrogen bond analysis reveals a decrease in the number of intermolecular water-water hydrogen bonds per water molecule on increasing salt concentration. Network analysis suggests that as the salt concentration increases, the hydrogen bonding network of water molecules is disrupted, along with the concurrent formation of a TFSI network. These solvation environments computed from the FPMD simulations were compared with the solvation environments for the same system size calculated from simulations using two molecular mechanics force fields (FF), and a good agreement between the results was obtained.

Furthermore, to probe the dynamics of the system, diffusion coefficients of the species and the van Hove functions (from FF-based simulations) for the Li-water and Li-TFSI interactions are computed. Lithium diffusion coefficients at varying salt concentrations are compared with the experimental results. The structure decorrelation timescales obtained from van Hove functions are compared with the diffusion timescales to understand the lithium transport mechanism through the electrolyte. These insights into the structure and dynamics of the aqueous LiTFSI systems provide an opportunity to design electrolytes for Li-ion batteries with enhanced performance and safety.

531b Tuning Molecular Rigidity to Modulate Lithium Coordination and Enhance Mobility in Ionic Liquids

Matthew Gebbie, John McAlpine

As lithium-ion battery deployment rapidly expands, increases in battery safety and sustainability are becoming critical. Currently, battery flammability is a growing problem, which is exacerbated by the high flammability of organic electrolytes. Ionic liquid-derived electrolytes offer safe, nonflammable alternatives, but the performance of ionic liquids as battery electrolytes has been hindered by poor lithium transport, which arises from formation of strong lithium-anion coordination complexes. For example, recent studies suggest that anion coordination to lithium can even drive formation of negative complexes in ionic liquids, which may cause lithium species to move opposite of the desired direction. In this talk, I will discuss our work on tuning ionic liquid nanostructures to suppress lithium-anion coordination via incorporation of rigid functional groups. By leveraging the unique self-assembly properties of rigid molecular clusters, we show that lithium coordination can be suppressed and organic cation mobilities can be minimized, resulting in dramatically enhanced lithium mobilities compared to the surrounding organic ion matrix. Specifically, we use studies of phase behavior, vibrational spectroscopy, and magnetic spectroscopy to demonstrate that rigid groups can be tuned to maintain high lithium solubility while concurrently minimizing lithium-anion coordination. Further, we show that lithium mobilities can be enhanced through rational incorporation of molecular packing mismatches. Our results provide a new paradigm for enhancing lithium mobility in ionic liquid-derived electrolytes by tuning self-assembly of cation functional groups to enhance organic cation-anion interactions, suppress lithium coordination, and enhance lithium mobility.

531c High Energy Density Picoliter Batteries for Colloidal Robots and State Machines

Ge Zhang, Jing Fan Yang, Volodymyr Koman, Sungyun Yang, Allan Brooks, Matthias Kuehne, Michael Strano

The recent interest in microscopic autonomous systems, including microrobots, colloidal state machines and smart dust has created a need for microscale energy storage and harvesting. However, macroscopic materials for energy storage have noted incompatibilities with micro-fabrication techniques, creating significant challenges to realizing microscale energy systems. Herein, we photolithographically pattern a microscale Zn/Pt/SU-8 system to generate the highest energy density microbattery at the picoliter (10^{-12} L) scale. The device scavenges ambient or solution dissolved oxygen for a Zn oxidation reaction, achieving an energy density ranging from 760 to 1070 Wh L^{-1} at scales below 100 μm lateral and 2 μm thickness in size. More than 10,000 devices per wafer can be released into solution as functional colloids with energy stored onboard. Within a volume of only 2 pL each, these microbatteries can deliver open circuit voltages of 1.16 V with total energies ranging from 5.5 ± 0.3 to 7.7 ± 1.0 μJ and a maximum power near 2.7 nW. We demonstrate that such systems can reliably power a micron-sized memristor circuit, providing access to non-volatile memory. We also cycle power to drive the reversible bending of microscale bimorph actuators at 0.05 Hz for mechanical functions of colloidal robots. Additional capabilities such as powering two distinct nanosensor types and a clock circuit are also demonstrated. The high energy density, low volume and simple configuration promise the mass fabrication and adoption of such picoliter Zn-air batteries for micron-scale, colloidal robotics with autonomous functions. To meet the demand of different micro-robotic tasks, we are actively developing more types of picoliter batteries using different materials (such as Mg and Ag_2O) and configurations.

531e Optimization and Surface Characterization of Porous Zinc Anodes for Zinc Metal Batteries

Matthew Powell, Shuya Wei

The current energy storage market is dominated by lithium-ion batteries (LIBs); however, LIBs are reaching their theoretical energy density. Researchers are looking into alternative solutions in different battery chemistries ranging from metal-sulfur batteries to metal-air batteries. One type of metal air battery that has peak interest in recent years is zinc air batteries (ZABs). Zinc-based batteries are considered as a replacement for LIBs due to their inherent safety, abundance, low cost, and its 2-electron transfer. However, zinc, like all other planar metals, has uneven plating cycles that cause dendrites to form leading to the battery shorting. Therefore, it is critical to improve plating and stripping of zinc ions during the charge and discharge process to enhance the cycle life of the battery.

In this work, we have successfully synthesized porous zinc electrodes (PZE) via a gel-binder method that can stably charge and discharge for over 800 hours at 1mA/cm^2 before showing signs of failure. We compared PZEs synthesized from small(60nm), intermediate(10μm), and large(150μm) zinc particles to determine which surface features are best suited to mitigate dendritic growth and to improve electrolyte stability. From Scanning electron microscopy (SEM) images seen in **Figure 1**, large PZE exhibits the best platting patterns of the three PZE examined. The zinc deposits on the large PZE shows a stable and flat morphology and does not form the hexagonal close-packed (HCP) crystal structure that is seen on planar zinc anodes. The deposits seem to melt onto the surface, promoting even plating. The intermediate PZE has an increased affinity to deposit onto the glass microfiber separator leading to a decrease of active material and causing instability during galvanostatic cycling. SEM images of both planar zinc and small PZE show HCP that are normal to the surface, which resulted in very poor electrochemical performance. As the particle size increases, the deposits transition from HCP crystals to flat metal deposits, increasing cyclic stability. Linear sweep voltammetry (LSV) was also performed to test the stability of the electrolyte. The large porous structure of the electrode improved the stability of the electrolyte by increasing the voltage by 300mVs before the electrolyte decomposed compared to planar zinc.

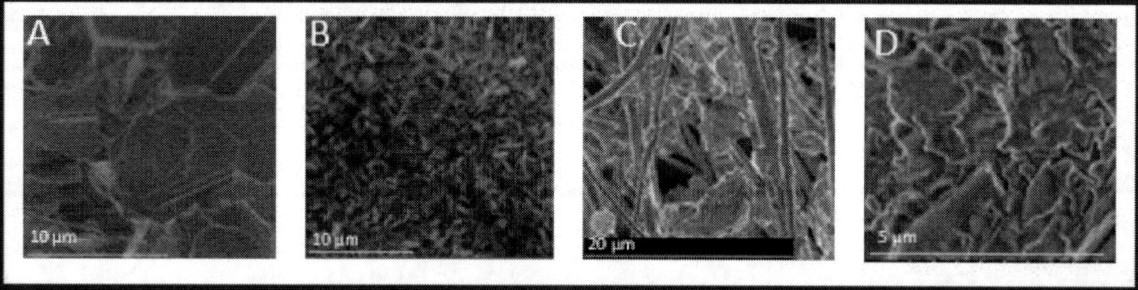

Figure 1: SEM images of shorted zinc anodes made from different particle sizes. Symmetric coin cells in 2M ZnSO4 + 5 wt% PEG300 were run under constant current conditions of 1mA/cm^2 until failure. (a)planar zinc (b) 60nm (c)10μm (d) 150μm.

531f Hierarchical Multiphase Porous Electrode Theory -- a Case Study of Porous Graphite Electrode

Huada Lian, Martin Z. Bazant

The porous graphite electrode is the most widely used anode in lithium-ion battery and its secondary structure such as porosity and specific surface area of particle can vary. In general, the specific surface area of a fresh graphite electrode is relatively low (~ 1 m^2/g) after coating and chemical treatment to reduce the capacity loss due to formation of solid-electrolyte interface(SEI). However, it still allows the electrolyte to fill up the nanopores and provides reactions on the internal surface of each secondary particle. The Damköhler numbers in both electrode scale and particle scale suggest that their concentration polarization are comparable and thus can not be ignored. Classic porous electrode theory assumes the active material to be compact and oversimplifies the flux of lithium ion in a secondary particle purely as a solid diffusion, therefore the diffusion coefficients extracted from experiments can sometimes vary several orders of magnitude. We combine the phase-field approach and a hierarchical porous electrode theory to capture the dynamics of reaction and transport in both the electrode scale and particle scale. The model is validated against the lithiation dynamics of porous graphite anode measured in-operando, capturing both the spatial and temporal nonuniformity. We then parameterize the model to a Li/Graphite half-cell setup and fit it into multiple data sets found in literature. The model respects the reaction kinetics and transport in a phase-separating material, thus transferable, and can guide the design of electrode and characterization methods for state of charge estimation and early detection of lithium plating.

531g Advanced Halogenated Electrolytes for Improved Performance, Safety, and Cycle Life of High Energy Lithium-Ion Batteries

Robert Emmett

High capacity cathodes are necessary for the development of future high energy density lithium-ion batteries, although challenges to safe charge transfer between these cathodes and the electrolytes need to be surmounted to create a cost effective battery solution. Stability of the solid-electrolyte interface layer is important for improving cycle life, necessary for most battery applications, and promoting safer operations, to prevent dangerous exothermic reactions that can cause thermal runaway. The Army is looking at these technologies as a way to increase the energy density of soldier worn batteries, which account for > 40 pounds of the ~130 pounds a soldier carries when deployed. In this project we utilized a in house developed chlorinated manganese spinel due to a > 200 mAh/g capacity and 5 V cell operation. Common high voltage electrolytes employ fluorinated carbonates that seem to cause a parasitic interaction between the chlorinated cathode materials and the electrolyte. Determination of the degradation effects will help inspire future electrolytes for chlorinated spinel materials for future battery technologies and create a > 400 Wh/Kg cell with stable cycle life. SEM, EDS, and XRD were utilized to evaluate the degradation of the cathode surface with repeated cycling. Additional, efforts are targeted in creating rapid charge transfer at the electrode / electrolyte surface with non-flammable electrolytes to further enhance the safety characteristics of these cells. Future high energy density batteries with stable charge transfer are necessary for the developing battery industry, especially electronic vehicles, and will help reduce the battery weight burden of soldiers while deployed on mission.

289a Multi-Scale Modeling of Novel Amine-Based Solvents for CO_2 Capture

Ismail Alkhatib, Daniel Bahamon, Lourdes Vega

Among the different available options for combating climate change and global warming, carbon capture and storage (CCS) is positioned as the most feasible for short- and medium-terms, with post-combustion capture (PCC) considered more suitable for direct integration with power generation facilities reliant on fossil fuels. Aqueous alkanolamines have been the long-standing benchmark solvents for this application, however, the limiting factor in their commercialization is the parasitic energy load associated with the regeneration of the solvent. This can be attributed to either the enthalpy of absorption associated with the chemical solvent, or the latent and sensible heat requirements associated with the physical co-solvent. Efforts over the past decades have been directed towards reducing the regeneration energy requirements by either proposing alternative novel amines with lower absorption enthalpy, or the use of alternative organic diluents with lower thermal properties than water, formulating water-free or water-lean solvents [1, 2].

However, demonstration of the potentiality of these solvents remains limited to lab-scale application, focused on a limited set of performance criteria such as absorption capacity, and energy of regeneration. Additionally, the limited availability of experimental data required for detailed process modeling and simulation, hinders the techno-economic evaluation of these solvents to accurately assess their potential over their aqueous counterparts. Therefore, devising a consistent and reliable thermodynamic model capable of accurately describing the CO_2 solubility, thermophysical and transport properties of alternative amine-based solvents for representative techno-economic evaluation is highly needed.

In this contribution, we demonstrate the application of a multi-scale modeling framework to assess the potentiality of alternative amine-based solvents for CO_2 capture examining novel aqueous amines, water-free, and water-lean solvents. The multi-scale framework is built in a consistent manner establishing a direct link between the molecular structure of the solvent and its performance on industry-scale. The core of the multi-scale modeling framework relies on the application of a robust molecular-based equation of state (EoS), namely, soft-SAFT EoS [1,2], to provide all required thermodynamic information needed for an accurate process model and techno-economic evaluation, such as CO_2 solubility, solvent viscosity, density, surface tension, *etc.* Investigated solvents include novel aqueous amines from novel families such as alkyl-ethanolamines, amino-alcohols, dialkylamino-alcohols, and multiamines. Additionally, water-free and water-lean solvents formulated from the chemical solvent monoethanolamine (MEA), with a wide array of organic diluents such as alcohols, glycols, glymes, and polar aprotic solvents were also examined. Molecular parameters of all the investigated substances are either taken from previous contributions or developed in this work and used in a transferable manner for describing the solubility of CO_2 in the selected solvents at conditions of relevance for industrial applications. The CO_2 chemisorption process in the examined solvents is done by using a scheme of implicit reactions to describe the formation of carbamate resulting from the chemical reactions between CO_2 and the examined amines [3,4]. This approach eliminates the need to specify detailed equilibrium reactions, while significantly reducing the number of parameters required to capture the CO_2 solubility, allowing the application of the model in a predictive manner over a broad range of conditions unavailable from experimental data.

Once these solvents were fully characterized using available experimental data, the thermodynamic model was used in a predictive manner to provide required thermodynamic information for detailed process model and simulation for a post-combustion capture unit. This enabled the assessment of the performance of these solvents in terms of technical indicators such as absorption capacity, and energy consumption, along with economic indicators inclusive of capital, operating costs, and total annualized costs, providing a direct link between micro-level features and industry-level performance [5].

This work is funded by Khalifa University of Science and Technology (RC2-2019-007). Computational resources from the Research and Innovation Center on CO_2 and H_2 (RICH Center) are gratefully acknowledged.

References

[1] F.J. Blas, L.F. Vega, Thermodynamic behaviour of homonuclear and heteronuclear Lennard-Jones chains with association sites from simulation and theory, Mol. Phys. 92 (1997) 135–150. https://doi.org/10.1080/002689797170707.

[2] F.J. Blas, L.F. Vega, Prediction of Binary and Ternary Diagrams Using the Statistical Associating Fluid Theory (SAFT) Equation of State, Ind. Eng. Chem. Res. 37 (1998) 660–674. https://doi.org/10.1021/ie970449+.

[3] L.M.C. Pereira, L.F. Vega, A systematic approach for the thermodynamic modelling of CO2-amine absorption process using molecular-based models, Appl. Energy. 232 (2018) 273–291. https://doi.org/10.1016/j.apenergy.2018.09.189.

[4] I.I.I. Alkhatib, L.M.C. Pereira, A. Alhajaj, L.F. Vega, Performance of non-aqueous amine hybrid solvents mixtures for CO2 capture: A study using a molecular-based model, J. CO2 Util. 35 (2020) 126–144. https://doi.org/10.1016/j.jcou.2019.09.010.

[5] H.A. Balogun, D. Bahamon, S. Almenhali, L.F. Vega, A. Alhajaj, Are we missing something when evaluating adsorbents for CO2 capture at the system level?, Energy Environ. Sci. 14 (2021) 6360–6380. https://doi.org/10.1039/D1EE01677F.

289b Multi-Scale Dynamic Modeling and Techno-Economic Analysis of a Radial Flow Fixed Bed Contactor for Post-Combustion CO_2 Capture

Ana Flavia Monteiro, Ryan Hughes, Debangsu Bhattacharyya

Solid sorbent contactors are a commonly used technology for post-combustion CO_2 capture as they can potentially reduce capital and operational costs of the process. While fluidized, rotary, or moving beds may increase mass and heat transfer efficiency, they can be difficult to operate, require an attrition resistant sorbent, and can be difficult to scale-up from experimental data[1]. Fixed bed contactors are easy to construct, operate, and run on a lab scale. Common fixed bed contactors have an axial flow configuration. This configuration, however, leads to large bed volumes and high pressure drops, especially when large volume of flue gas with low concentration of CO_2 needs to be treated[2]. Radial flow fixed bed contactors (RFBR) can mitigate some of the problems mentioned above for the other type of contactors. In an RFBR, the gas flows radially in the reactor, which allows for a higher surface area for gas-solid contact, reducing the size of the vessel and the pressure drop, and therefore the cost of the process. RFFB contactors have been investigated for many applications, including direct air capture[3, 4], oxygen generation[5], and various pressure swing adsorption processes[5], but a detailed evaluation of RFFB contactors for post-combustion CO_2 capture using temperature swing adsorption that can be used for scale-up and optimization of a commercial system is lacking.

In this work, an RFBR model was developed for a functionalized metal organic framework (MOF). The MOF exhibits an unusual step-shaped isotherm that make it challenging to develop an isotherm model. A model of the sorbent isotherm is developed and validated by using the experimental data. The kinetic model is developed based on the thermogravimetric analysis data. In addition, a multi-scale model is developed. A particle level model is developed that incorporates mass and heat transfer resistance. This model is integrated with a bulk scale model. The multi-scale model is validated with the experimental breakthrough data from a lab-scale reactor. The model is then scaled up for a commercial scale simulating temperature swing adsorption cycle. For this MOF, effective thermal management is of high importance for realizing the potential of this MOF. An economic model was developed and techno-economic analysis was carried out by simulating changes in a number of design and operating variables. Results are compared with the axial flow fixed bed contactor showing considerable differences not only in the pressure drop and volume as has been reported in the literature but also the economics. In particular, capital and operating cost differences between these configurations are of utmost importance for commercialization of these technologies.

.

References:

[1] Menendez, M., Herguido, J., Berard, A., Patience, G., Experimental Methods in Chemical Engineering: Reactors – Fluidized Beds. *The Canadian Journal of Chemical Engineering*. 2019; 97, 2383-2394.

[2] Kareeri, A.A., Zughbi, H.D., and Al-Ali, H.H., Simulation of Flow Distribution in Radial Flow Reactors. *Ind. Eng. Chem. Res.* 2006; *45*, 2862-2874

[3] Yu, Q., Brilman, W., A Radial Flow Contactor for Ambient Air CO2 Capture. *Appl. Sci.* 2020; 10, 1080

[4] Schellevis, M., Jacobs, T., and Brilman, W., CO2 Capture From Air in a Radial Flow Contactor: Batch or Continuous Operation?. *Frontiers in Chemical Engineering.* 2020*;* 2.

[5] Wang, H., Yang, X., Li, Z., Liu, Y., Zhang, C., Xiaojun Ma, X., and Chunwang Li, C., 3-D Modeling of Gas–Solid Two-Phase Flow in a π-Shaped Centripetal Radial Flow Adsorber. *Appl. Sci.* 2020, 10, 614.

289c Development of a Novel Cu-Based Dual Function Material for CO_2 Capture and Conversion

Rim Ismail, Mohammed Ali Saleh, Abdelbaki Benamor, Sardar Ali, Muftah El-Naas

The industrial revolution and energy demand have significantly increased the greenhouse gas emissions, among them CO_2, which is responsible for severe environmental concerns. To this end, various CO_2 mitigation systems have been sought, with more research has focused on integrating CO_2 capture and utilization into valuable products. Recently, dual functional materials (DFMs) that couple adsorbent and catalytic species for CO_2 capture and conversion are notably developed as promising solution. In this study, an investigation of the catalytic performance of Cu-CaO and Rh-promoted Cu-CaO supported over high surface area alumina, for CO_2 conversion to high value products is presented. The results indicated that Cu loading of 10%, 20%, and 30% impregnated onto 10%CaO/Al_2O_3 exhibited higher activity and attained CO_2 conversion of 15%, 15.4%, and 16.44%, respectively at 300 ^{0}C and 60 bar. The effect of adsorbent composition, promotor addition and process variables were tested for a selected catalyst where the highest CO_2 conversion and MeOH yield were observed at about 350 ^{0}C, 80 bar, and 5000 h^{-1} GHSV.

289d High Surface Area Guanidine Based Sorbents for CO_2 Capture

Mani Modayil Korah, Matthew D. Green

Global warming is undoubtedly one of the greatest challenges facing mankind in the 21st century. The accumulation of CO_2 in the atmosphere has been documented to be the primary driver of global warming. The global average CO_2 levels in the atmosphere have gone up from 340 ppm in the 1980s to 415 ppm in June 2021. The 2021 report from the Intergovernmental Panel on Climate Change (IPCC) predicts a 1.5 °C rise in global temperatures within the next two decades. This scenario assumes that global policymakers take an aggressive stance on cutting down emissions. Under a high emissions scenario, the IPCC report predicts that the world may warm up to 4.4 °C, leading to catastrophic results. With current CO_2 emissions set to reach 40 GT CO_2/year, the capture and storage of CO_2 is essential to mitigate climate change. Point source CO_2 capture alone from emission sources like power plants, refineries, and factories is no longer sufficient since it does not address CO_2 emissions from dispersed sources such as the transportation sector, housing, etc. Non-point sources account for ~50% of global CO_2 emissions and, thus, technologies that capture CO_2 from ambient air are needed to achieve net-zero emissions globally by 2050. Returning to preindustrial CO_2 levels and keeping global temperatures below the target of 1.5 °C set by the Paris Climate Accord necessitate negative emission technologies, and direct air capture (DAC) is the only technological solution capable of achieving that goal.

DAC, while nascent, is a promising technology that offers a path to negative CO_2 emissions and paves a path to remediating the anthropogenic impact on climate. DAC can complement point-source CO_2 capture and storage (CCS), compensate for residual emissions that escape during point-source capture, and target distributed emissions. Traditionally, liquid sorbents like aqueous solutions containing potassium or sodium carbonate/hydroxide or alkanolamine solvents like monoethanolamine (MEA) or diethanolamine (DEA) have been used to capture CO_2. Reasons including slow kinetics, loss of sorbent via evaporation, high regeneration temperatures hinder commercial viability of such technologies. Guanidine with its high amine availability has become a molecule of interest in recent years for CO_2 sorption. In this work, we explore high surface area guanidine based polymeric sorbents that serve as high performance materials with excellent kinetics, sorption capacity, and fast and low-energy regeneration.

First, we examine the CO_2 sorption potential of L-Arginine amino acid that has a guanidine group in the molecule. We show that the incorporation of L-Arginine onto high surface area poly(vinyl alcohol) (PVA) electrospun mats significantly improves sorption performance. Amino acids like L-Arginine, have the same functional group as alkanolamines (e.g., primary amines, and some also contain alcohol groups) and are potentially environmentally friendly. Amino acid solutions also have higher surface tension, lower volatility, better chemical and thermal stability, and exhibit similar or better CO_2 absorption compared to alkanolamines. L-arginine (L-Arg), in particular, has shown excellent CO_2 sorption capacity. By incorporating the amino acid L-arginine (L-Arg) onto a poly(vinyl alcohol) (PVA) nanofiber support, we created porous substrates with very high surface areas of L-Arg available for the sorption of CO_2. The PVA-Arg nanofiber composites are flexible and show excellent passive DAC performance compared to bulk L-Arg. The nanofiber composites are fabricated via an electrospinning process using an aqueous polymer solution. After obtaining the nanofiber mats, they are physically crosslinked to improve resistance to water. Increasing L-Arg loading to 50 wt.% (with respect to PVA) improves sorption performance; however, loadings above 50 wt.% decrease the performance due to a reduction in porosity of the nanofiber composite. Higher ambient humidity levels also improve sorption performance significantly. The best performing nanofiber composite shows a maximum adsorption capacity of 542 $\mu mol\ g^{-1}$ following 2 h of exposure to ambient, high humidity air that had a CO_2 concentration of 445 ppm.

Secondly, we demonstrate the potential of poly(vinyl guanidine) (PVG) as a potential DAC sorbent. PVG is synthesized in three steps, starting with the free radical polymerization of vinyl formamide into poly(vinyl formamide). Poly(vinyl formamide) is then hydrolyzed using NaOH into poly(vinyl amine). The guanidine groups are introduced into the polymer via reaction with cyanamide under acidic conditions. PVG crosslinked with Gluteraldehyde (GA) is also synthesized to make the sorbent water resistant. Both the PVG and the GA crosslinked PVG show excellent DAC capabilities. Electrospinning the PVG into nanofibers further helps the material to attain fast kinetics and high sorption capacity on exposure to lab air. The PVG nanofibers also show excellent passive CO_2 sorption capacities (attached Figure) of upto 703 μmol g^{-1} over a period of 2h while also displaying fast desorption kinetics. We further explore the role of form factor in improving the kinetics of sorption for PVG sorbents.

To simulate real world conditions, all testing of our sorbents were performed using lab air, simulating average wind speed conditions, with no packing of the material in the testing chamber. Thus, we successfully demonstrate passive CO_2 sorption of these high surface area materials. Passive direct air capture of CO_2 is required to make the process energy efficient and commercially viable. Thus, we successfully demonstrate the capabilities of high surface area guanidine polymeric materials for potential use in real world DAC applications.

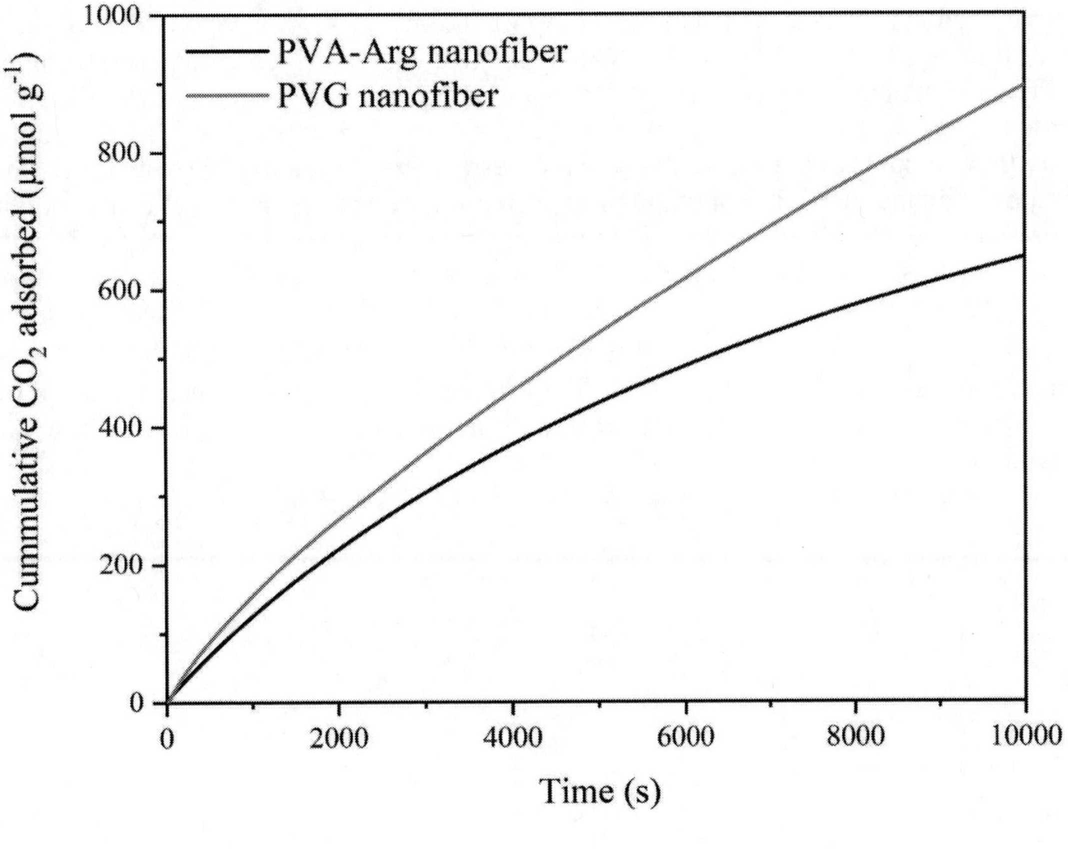

289f A Passive Wind Collector Integrated with a Direct Air Capture (DAC) System for Efficient, Sustainable, and Scalable CO_2 capture

James Akinjide, Joo-Youp Lee, Aashish Priye

One of the major challenges facing our current CO_2 mitigation technologies is the scale-up of direct air capture (DAC) systems at the large scale to meet the desired target goal of <$100/tonne CO_2 removed. Conventional DAC systems rely on fans to drive the ambient air through a CO_2 adsorption system to produce air with lean CO_2 levels. It is imperative to passively collect and drive wind to reduce a continuous operating cost (i.e. fan) required to introduce air flow to DAC system while providing good mass transfer for subsequent CO_2 capture in the system. Here we introduce a wind collector that can passively collect winds from all directions and drives it into a CO_2 adsorption system containing sorbent coated monolithic structure with open cells while still maintaining low pressure drop across the adsorption unit to ensure high operational throughput.

Even though the concept is simple, the overall efficiency of CO_2 adsorption in such a passive DAC system depend on the intricate and coupled effects of (i) available wind velocities (ii) pressure and velocity distribution throughout the DAC system and (iii) mass transfer and adsorption characteristics of the sorbent coated monolith. To resolve this complex interplay, we developed a comprehensive computational fluid dynamics (CFD) model to analyze the effect of critical geometric parameters of the wind collector (opening width (R), height of side opening (H), and opening angle (θ)) and the monolith structure (pitch size (P), Length (L)) on the transport of air through the open monolith cells. Depending on the pressure drop generated inside the passive DAC system (due to internal flow resistance imposed by the monolith), a range of expected internal air velocities at different wind speeds can be achieved. Results from the CFD model provide us with valuable insights into the dependence of system's geometric parameters on the CO_2 capture efficiencies and location specific optimal design parameters based on the locally available wind velocities. As there are no moving parts in this design, the maintenance requirement is very low. The high throughput removal of CO_2 from ambient air in a system that offers a small footprint, scalability, and off grid operation (no electricity required) has the potential for widespread implementation, enabling us to address the CO_2 mitigation needs at large scales.

289g Novel Facilitated Transport Membrane for Low Cost Carbon Capture in Cement, Steel, and Steam Methane Reforming

Christine Parrish, Ken Loprete, Sudip Majumdar

With 40 billion tons of anthropogenic green house gases (GHG) being added to the atmosphere every year, it is vital to make existing processes carbon neutral to prevent catastrophic climate change. A majority of the world still relies on fossil fuel technology for power and to provide the chemical building blocks for today's materials. Carbon capture technology allows users to filter and sequester the CO_2 at the source and reduce emissions by up to 90%. Currently, separation and concentration of the CO2 stream in preparation for liquefaction and storage is done using amine towers. These towers are costly (>$50/ton of CO2) and have a parasitic energy load of 20-40%, making them unfeasible for wide adoption. Hard to abate industrial sectors like cement, steel, and grey hydrogen are unlikely to be served by electrification but are equally as unlikely to be replaced by novel technology. Compact Membrane Systems (CMS) has developed a novel facilitated transport membrane to deliver low cost carbon capture. Three use cases will be discussed: cement, steel, and steam methane reforming.

Optiperm™ *carbon* is a facilitated transport membrane with significantly higher permeance than existing membrane technology. The technology uses a fluoropolymer composite membrane with a facilitating agent incorporated into the polymer to pull CO2 across the membrane. The increased flux and high selectivity positions Optiperm™ *carbon* to reduce the cost of carbon capture across multiple applications. Membranes operate in energy efficient modes, so can reduce the parasitic load imposed by alternative technologies like amine towers. Results thus far have shown a very high flux membrane (2000-5000 GPU) with adequate selectivity (18-30) at realistic conditions that would result in substantially lower capture costs and increased efficiency. The membranes attain this performance at elevated temperature (60-100°C) and in the presence of poisons (SOx and NOx). Results from the aging and poison testing will be discussed.

Because Optiperm™ *carbon* has a unique set of characteristics. CMS has been able to design a two stage cascade approach to decrease the energy use while maintaining 90% CO2 recovery and 95% CO2 product purity. Design considerations and cost sensitivities for operating conditions will be discussed fot the three use cases.

540a Hybrid Approach for Accelerating the Implementation of Hydrogen Infrastructure Using Experience Based Data Formatting and Systematic Decision-Making Framework

Soo Hwan Kim, Jun-Hyung Ryu

Energy plays the pivotal role over entire economic activities. Huge cost with long times should be consumed in developing and implementing an energy system based on an energy source. The process of 'developing' involves tackling various technical breakthroughs in commercializing the corresponding energy system. The implementing one involves constructing a large number of facilities over wide areas which are often geographically scattered. It costs more severe time and cost to shift from one energy source into another.

The current world is under such tremendous pressure of shifting from fossil fuel based energy system into a new one. Renewable energy source is definitely preferred but there are sectors in which they cannot meet the entire demand. Transportation sector is a case where hydrogen is emerging as a promising alternative. There are many advantages in employing hydrogen as an energy carrier in the sector such as environmentally clean emission gas. Efforts have been conducted over the world to accelerate the implementation of hydrogen infrastructure.

It is not easy to construct hydrogen infrastructure. Its production cost is still so high that it is not fully competitive against fossil fuels. Furthermore, a small number of hydrogen refueling stations (HRS) still poses handicap that makes the public hesitate to purchase hydrogen vehicles.

This paper introduces how past experiences of Process Systems Engineering (PSE) community can be used to tackle issues in accelerating the hydrogen infrastructure. Particularly hybrid approach in terms of systematic decision-making framework with data preparation process by using data-driven formatting. Case studies illustrating the construction of HRSs in South Korea are presented to illustrate the applicability of the proposed framework.

540b Supercapacitors Based on Graphene Quantum Dots for Efficient Energy Storage

Ayse Saliha Korkut, Betul Uralcan

Electrical double layer capacitors (EDLCs) store energy through physical adsorption of ions as a response to an applied potential difference. As their energy storage mechanism relies on physical interactions, EDLCs have high power densities, unlike batteries, which are often limited by the slow charge and mass transfer kinetics. Additionally, EDLCs can sustain millions of charging/discharging cycles. Nevertheless, they suffer from low energy densities. For a high energy density EDLC, it is critical to maintain a compact architecture with large ion-accessible surface area while also ensuring low ion transport and electrical resistance. We incorporate carbon quantum dots into thermally exfoliated graphene oxide sheets in the presence of a room temperature ionic liquid to form conductive carbon networks, enhancing ion transport kinetics and storage. This yields an electrode in which both carbon quantum dots and ionic liquid serve as spacers to effectively separate the thermally exfoliated graphene oxide sheets, while ionic liquid also functions as electrolyte. These findings demonstrate both energy and power density can be improved simultaneously by tailoring electrolyte composition and carbon quantum dot/thermally exfoliated graphene oxide ratio.

540c Optimization of Energy Density in Supercapacitors By Utilizing a Hybrid Artificial Neural Networks-Genetic Algorithm Based Optimization Algorithm

Duygu Kaya, Betul Uralcan

Supercapacitors are promising energy storage devices with long cycle life and high-power density. Nonetheless, they have limited energy density. In this work, we utilize a hybrid Artificial Neural Network-Genetic Algorithm (ANN-GA) model to optimize the energy density of supercapacitors while maintaining their high power density. We extract 2370 data points from the literature on carbon-based supercapacitors, and analyze the dataset using simple descriptive statistics to show the general trends in the field. Then, we perform preprocessing and feature selection, and apply machine learning algorithms to predict energy storage performance. Artificial neural networks yield the best accuracy. Consequently, we use the trained neural network as the fitness function for genetic algorithm to identify a set of optimal parameters that yield improved energy and power performance. Our findings indicate that obtained optimal design parameters are consistent with the literature. In a nut shell, this study takes a step towards the rational design of supercapacitors by implementing a hybrid ANN-GA as an optimization tool to improve energy storage performance.

540d Tuning Gas Diffusion Electrode Construction to Mitigate Flooding and Degradation Processes for Electrochemical CO2 Reduction

Khantey Lim, Xiao Kun Lu, Bingzhang Lu, Linsey Seitz

A critical limitation of overall electrochemical CO_2 reduction (ECR) performance is suboptimal construction of gas diffusion electrodes (GDEs). GDEs separate the gaseous CO_2 reactant feed from the liquid electrolyte, thereby enhancing mass transfer in an otherwise solubility-limited electrocatalytic reaction and enabling efficient operation at high current densities. However, commercialized GDEs starts to lose their hydrophobicity after ~2hrs of electrocatalytic reaction, leading to flooding and failure of the entire reactor system. Compared to catalyst development, there is scant work towards fundamentally understanding GDE's local reaction environment and the process of electrode degradation. Herein, we prepare 2 different catalysts (Cu and CuO) with varied loadings of PTFE particles and carbon black deposited on PTFE membrane or carbon paper supports to investigate overall GDE properties and monitor time-dependent changes. We alternate ECR reaction with electrochemical impedance spectroscopy (EIS) and goniometer contact angle measurements to track changes in impedance and hydrophobicity as a function of GDE construction and operating conditions (i.e. time, applied potential). EIS data is fit to appropriate equivalent circuit models to provide insights regarding possible charge transfer mechanisms for the system. We analyze contact angle measurements using the Cassie Baxter equation to explicitly quantify the fractional surface area of gas-dominated regions that are caused by the varied PTFE loading and ECR operating conditions. We find that GDE made from PTFE membrane/spray casted carbon black/catalyst to be incompatible for further analysis at different time intervals as the PTFE membrane is innately insulating and leads to high ohmic loss. Low carbon loading leads to high series resistance ($>20\ \Omega$) while higher carbon loading favors undesirable hydrogen evolution reaction (HER). Results from this work will guide continuing investigation and improved GDE construction to mitigate degradation processes.

375f Construction of a Sensor for Rapid and Online Measurement of Dissolved Methane in Bioreactor Systems

Robert Bertrand, Lisa Stephanie Dizon, Rafael Hernandez, Mark Zappi, Emmanuel Revellame

Two phase partitioning bioreactors (TPPBs) have seen a rapid increase in study over the past several years due to their usage with methanotrophs to remediate methane emissions and generate useful products. In order to use these reactors effectively, the ability of the reactor to promote the gas transfer of methane into the liquid phase must be independently determined, which is something that few studies have tackled in recent years. The objective of this study was to construct and test a low-cost and easily implemented dissolved methane sensor for the estimation of the gas transfer coefficient of methane in TPPB reactors, which have been shown to selectively promote methane transfer into the media. Much of the research in the field currently relies on correlations between the oxygen transfer (a much easier measurement to take) and the methane transfer, which is simply not viable for TPPB reactors. This study is based on previous work [1] to independently measure dissolved methane, and expands on the capabilities for integration and data collection, while providing more knowledge and plans for future researchers. The principle of operation is that methane can preferentially dissolve through a thin silicone membrane between liquid and gas, causing dissolved methane to migrate into the gas phase on the opposite side of the membrane. If the other side of the membrane is flushed with a purge gas, this dissolution may be promoted by maximizing the concentration gradient between the two phases. The gas may then be passed through a traditional gas-phase methane sensor (which are highly selective, sensitive, and cheap) for determination of the liquid-phase dissolved methane concentration. The rate of diffusion across this membrane (and thus the concentration in the purge gas) is correlated to the concentration The silicone tube used for testing was 10cm in length, with a 0.25 cm and 0.35 cm inside and outside diameter respectively (Detakta Co. PN: 02502). The length of the tube was 10cm Based on previous studies, the nitrogen purge gas flowrate was specified at 50 mL/min. This flowrate provides the best combination of sensitivity (negatively impacted by increasing gas flowrate) and rapidity (positively impacted by same). A housing was designed in Fusion 360 and 3D printed to direct the purge gas flow into the mesh dome of the sensing circuitry. The primary objective of this study is to produce an easily replicated sensor for measuring the volumetric mass transfer rate of methane gas, which is an important factor in bioreactor operation, design, and scale-up. Measurements of this parameter were performed using the gassing-out method, in which the sensor is acclimated with a purge of pure nitrogen (to remove any trace methane), and then sparged with methane (in this case, 100% methane gas at 2 L/min). The response of the sensor as it reaches a new equilibrium at saturation is linearized to yield the value of the transfer coefficient. The response over time of the sensor is demonstrative of its effectiveness even under changing gas conditions. For instance, it appears to have (as stated) a high selectivity for hydrocarbon gases, with no change in the baseline reading of the sensor when nitrogen, air, or helium are passed through the liquid phase. The two phase partitioning bioreactor (TPPB) used for the evaluation of the sensor had a volumetric mass transfer coefficient of 37.8±2.2 hr-1, which is much higher than that of previous studies, which show around 15 hr-1 [2]. This is credited to the oil circulation of the bioreactor, which promotes the gas transfer from the methane gas bubbles to the aqueous phase. Little concrete research exists, however, for the measurement of the methane transfer coefficient in bioreactors. This is likely due to the difficulty in constructing such a sensor, and many works use a correlation to relate the transfer rate to that of oxygen, a much easier measurement to achieve. This construction and validation lays the groundwork for an independently calculated methane transfer coefficient, a necessary element for the progression of methanotrophic bioreactor research, especially for TPPB style reactors.

423b Going Faster: Analytical Models for Predicting Cell Cycling Performance in Redox Flow Batteries

Bertrand J. Neyhouse, Jonathan Lee, Fikile R. Brushett

Abstract

Global decarbonization of the energy sector necessitates development of storage technologies to mediate the inherent intermittency of renewable resources. Electrochemical systems are well-positioned to support this transition, with redox flow batteries (RFBs) emerging as a promising platform due to their scalability, simplified manufacturing, and long service lifetimes.[1] However, current embodiments are too expensive for broad adoption, as they rely on costly charge-storage species and reactor components, motivating investigations into alternative material sets. For example, molecular engineering offers an expansive design space for redox species, enabling a myriad of potential combinations.[2] Despite this versatility, next-generation materials must carefully balance complex tradeoffs between power / energy density, cycling stability, energy efficiency, and capital costs. Navigating this multifaceted parameter space requires suitable models to unambiguously connect underlying constituent material properties to cell performance metrics.

Numerous modeling efforts of varying degrees of complexity and detail have sought to assess the electrochemical and fluid dynamic processes that govern cell behavior, providing insight into features which govern capacity fade and cycling efficiencies for certain systems. In particular, zero-dimensional models are typically used to describe time-dependent changes in redox species concentrations while ignoring spatial variations,[3–5] as such frameworks can more easily facilitate simulations of durational cycling (tens to hundreds of cycles), which are necessary to understand the long-term impact of parasitic losses (e.g., crossover, species decay). However, these approaches still rely on numerical solutions to the governing mass balances, making them computationally inefficient for multi-cycle simulations and thus challenging their use in broader parametric studies. By instead developing analytical models, we can markedly reduce computation times and enable analyses that were previously unachievable for systematic RFB diagnostics.

In this presentation, we introduce a zero-dimensional modeling framework to describe galvanostatic cycling in redox flow cells. Using a generalized set of constitutive equations, we derive analytical solutions for mass balances, charge / discharge behavior, and performance metrics (e.g., capacity fade, energy efficiency). To illustrate their utility, we explore several representative scenarios that highlight key relationships between material properties, operating conditions, and cycling performance. Importantly, the rapid computation times associated with analytical models facilitate their integration into more complex mathematical schemes; to this end, we discuss applications of this framework in techno-economic assessments, process control and optimization, parameter estimation from experimental studies, and synthetic data generation. Ultimately, this modeling toolkit provides a platform for exploring the impact of specific cell components whose properties govern multiple processes, helping to elucidate key performance descriptors and to identify favorable materials combinations for targeted energy storage applications.

Acknowledgments

This work was supported by the Joint Center for Energy Storage Research, an Energy Innovation Hub funded by the U.S. Department of Energy, Office of Science, Basic Energy Sciences. B.J.N gratefully acknowledges the NSF Graduate Research Fellowship Program under Grant Number 1122374. J.L gratefully acknowledges support from the MIT Materials Research Laboratory REU Program, as part of the MRSEC Program of the NSF under grant number DMR-14-19807. Any opinion, findings, and conclusions or recommendations expressed in this material are those of the authors and do not necessarily reflect the views of the NSF.

References

1. M. L. Perry and A. Z. Weber, *J. Electrochem. Soc.*, **163**, A5064–A5067 (2016).
2. Y. Ding, C. Zhang, L. Zhang, Y. Zhou, and G. Yu, *Chem. Soc. Rev.*, **47**, 69–103 (2018).
3. M. Pugach, M. Kondratenko, S. Briola, and A. Bischi, *Applied Energy*, **226**, 560–569 (2018).
4. S. Modak and D. G. Kwabi, *J. Electrochem. Soc.*, **168**, 080528 (2021).
5. B. J. Neyhouse, A. M. Fenton Jr., and F. R. Brushett, *J. Electrochem. Soc.*, **168**, 050501 (2021).

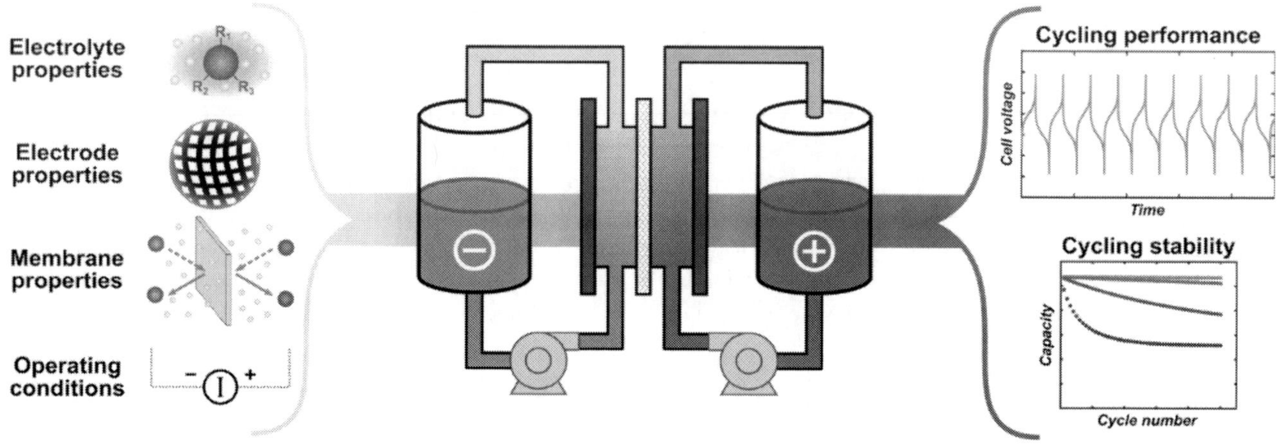

423c Anode Lithium Plating Detection and Charging Rate Optimization in Lithium-Ion Batteries

Arun Muthukkumaran, Raghunathan Rengaswamy

Lithium-ion batteries have played a crucial role in enabling the digital revolution. The technology has established itself as a reliable energy storage mechanism over the past decade. But the technology still suffers from problems such as safety issues, extended charging time, and degradation with cycling. One of the major concerns of safety is the lithium plating phenomenon encountered in lithium-ion batteries. Lithium plating is the reduction of lithium ions on the surface of the negative electrode while charging a lithium-ion battery. Lithium plating reaction can eventually lead to dendrite formation and shorting of the battery. The lithium plating phenomenon has been attributed to two major factors viz. increased charging voltage and charging at low temperatures. Charging a battery under these conditions leads to increased concentration of lithium-ions at the negative electrode than what can be intercalated into the electrode. This leads to the ions getting reduced on the surface of the electrode when the potential of the electrode falls below 0V.

In order to understand the degradation of the lithium-ion battery at different temperatures a first-principles based electro-thermal battery model was developed. The model incorporates two different parasitic side reactions viz. lithium plating reaction and the solid electrolyte interphase (SEI) layer formation reaction. These two reactions consume the lithium-ions present in the battery which in turn leads to degradation of the battery. In the first principles-based model the rate constants associated with the desired intercalation reaction as well as the undesired side reactions are coupled with the temperature of the cell. Hence, the model is able to detect changes in desired and undesired reactions at different temperatures. The model is also able to quantify the amount of lithium lost to the above-mentioned side reactions.

Charging rate is a user-controlled parameter which is gaining increased attention because of the requirement to fast charge the battery. However, fast charging by increasing the charging rate or charging voltage can also accelerate the degradation reactions inside the battery thereby reducing the life of the battery. This might also create safety hazards. The temperature of the battery plays a significant role in determining the extent of degradation at different charging rates and charging voltages since the rates of these reactions are dependent on temperature. There is a need to arrive at an optimal condition for charging taking into consideration all the above-mentioned parameters. Towards this end, a statistical optimization approach will be employed to identify optimal charging conditions at different ambient conditions of the battery. The objective of the optimization is to minimize the amount of lithium lost to plating and SEI layer formation. This approach which relies on understanding the degradation of the battery with the operating temperature will help in reducing the degradation of the battery while improving its safety.

474a Molecular-Level Understanding of Phase Stability in Model Phase-Change Nano-Emulsions for Thermal Energy Storage Investigated By NMR Spectroscopy

Jungeun Park, Ulrich Scheler, Robert Messinger

Phase Change Materials (PCMs) are latent heat storage materials that can store or release large quantities of energy while undergoing thermodynamic phase transitions. Organic PCMs have been great interest of large-scale applications since they are low cost, melt congruently, and exhibit good nucleating properties. Organic PCMs can be emulsified in water in the presence of surfactant to enhance thermal conductivity significantly as well as the transport property. However, PCMs nano-emulsions often become unstable during thermal cycling due to repeated melting and freezing cycles, which can further be affected by shear flow during their applications in heat transfer system. However, the distribution and dynamics of surfactant upon thermal cycling and shear are poorly understood, but are expected to correlate with observed phase, rheological and heat transfer instabilities.

To better understand the molecular origins of stability in PCMs nano-emulsions, a model PCM nano-emulsion system was synthesized using octadecane as the oil phase, stearic acid as the surfactant, and dilute aqueous NaOH as the medium. Solution-state nuclear magnetic resonance (NMR) methods have been applied to a PCM nano-emulsion in the presence of ^{13}C-enriched (50% uniformly ^{13}C-labeled) stearic acid under thermal cycling. Freezing and melting of ^{1}H octadecane nano-emulsion were investigated quantitatively, yielding information on octadecane supercooling and the content of liquid and oil within octadecane. Quantitative single-pulse ^{13}C NMR methods applied to show that the ^{13}C carbonyl head groups of the surfactants were present in multiple local environments and their signal intensities decreased upon thermal cycling thus losing molecular mobility (>100 cycles). Rheo-NMR and velocimetry methods were applied to measure the temperature effects on the flow of PCM nano-emulsion compared to the flow of water at room temperature, at different shear rates. As the octadecane changes phase from solid to liquid, the shear banding disappears and the velocity profile returns to near-Newtonian behavior. We analyze the shear banding effects in terms of a mechanical instability due to emulsion aggregation, which changes the local viscosity. NMR chemical shift, relaxation, and flow behavior of the component species were investigated as indicators of phase stability. Overall, the results lay the groundwork towards molecular-level understanding of phase stability under thermal cycling and shear in PCM nano-emulsions for thermal energy storage.

474b Development of Phase Change Material for Low Temperature Based Thermal Energy Storage Application

Kamal Nayan

The present work is concerned with development of novel phase change material (PCM) out of fatty acid and paraffin wax. The burgeoning need for energy, along with growing environmental concerns, has hastened the development of renewable energy sources, particularly materials for energy storage. One of the most promising areas of study is the use of PCM for thermal energy storage (TES). The author in this study has developed a novel eutectic mixture out of fatty acid and paraffin wax. Differential scanning calorimetry (DSC) was used to examine the thermophysical parameters of the generated eutectic mixture. DSC findings showed that the developed eutectic mixture had a congruent melting point, as well as a suitable melting point range and latent heat of fusion, indicating that it can be employed as a PCM for various latent heat thermal energy storage applications. The accelerated thermal cycle test revealed no variation in the thermophysical behavior of the developed eutectic mixture, showing that the developed PCM is thermally stable up to 500 heating/cooling cycles. The FTIR analysis proved its chemical stability over 500 thermal cycles. The effect of varied heating and cooling speeds was also investigated. And the results showed that changing the scanning rate had no effect on the melting point and latent heat of fusion of the produced PCM. Economic analysis suggested that our developed PCM is cheap and readily available in market. And if its mass production is done, its cost can be further reduced by 30-35%. Its application in buildings, photovoltaic systems and therapeutic beds has also been discussed.

354a Novel Electrode Structures for Enhanced Transport

Jacob Spendelow

Porous electrodes are a key component in fuel cells and other energy technologies, but the random and chaotic nature of state-of-the-art electrode structures leads to slow transport and reaction within the electrodes. Development of improved electrode structures is needed to accelerate fuel cell deployment. Advanced electrode designs can reduce barriers to transport and reaction, enabling fuel cells and related devices with higher power density and efficiency. Our team has developed new electrode architectures that provide enhancements in transport of both protons and oxygen using dedicated transport channels, non-tortuous geometries, and hierarchical structural control. These rationally-designed electrodes provide higher performance and durability than conventional electrodes in fuel cell membrane electrode assembly testing. The effect of structural parameters on electrode behavior will be discussed, along with outlook for future electrode designs.

354b Improving Performance and Durability of Polymer Electrolyte Fuel Cell Cathodes with High Oxygen Permeability Ionomer

Shawn Litster

This talk will cover recent advances in the understanding of the impact of the polymer electrolyte binder (i.e., ionomer) on the performance and durability of polymer electrolyte fuel cells for medium and heavy-duty vehicles. Ionomer is critical for proton current conduction through the porous catalyst layers as well as providing an acidic environment at the catalyst interface. However, there are several negative impacts of the thin ionomer coatings on platinum and platinum-alloy catalysts, including greater oxygen transport resistance and poisoning by the ionomer. The increased oxygen transport resistance due to the ionomer will be covered, including the effects of diffusion, solubility, and polymer chain densification at the catalyst surface. The second half of the presentation will focus on recent advances in the development and application of high oxygen permeability ionomer (HOPI) for cathodes. The talk will cover recent characterization, fabrication, testing, and modeling work, highlighting demonstrated reductions in the oxygen transport resistance as well as significant increases in efficiency, power density, and durability.

This work was partially supported by the U.S. Department of Energy, Office of Energy Efficiency and Renewable Energy under grant DE-0008822.

354c Recent Advances in Electrospun Fiber Mat Electrode MEAs for Hydrogen/Air Fuel Cells

Xiaozong Fan, Xiaomin Xu, Narae Kang, R. Wycisk, Peter Pintauro

When MEAs are made by a decal, catalyst-coated-membrane, or catalyst-coated GDL method, there is little or no control over the macro-scale organization of the catalyst and binder. Thus, the high ORR catalytic activity of Pt-based catalysts seen in rotating disk experiments is rarely observed in an operating fuel cell MEA. Features such as particle and binder interconnectivity, macroporosity, and microporosity become more critical when high-performance nanomaterial catalysts are employed in fuel cell electrodes. Consequently, new electrode fabrication techniques are needed for next-generation MEAs. Nanofiber electrospinning is just such a method, which can address fuel cell electrode manufacturing and structure/performance issues. As a fabrication method, electrospinning is scalable, robust, and cost-effective, especially for the creation of non-woven mats of sub-micron diameter polymer fibers. Although not as well studied, the method can also be used to prepare particle/polymer fiber networks with high particle loading and high intra- and inter-fiber porosity. Mats from such fibrous materials can be used as electrodes in fuel cells [1-5] where high interfacial electrode area is of prime importance. In this talk, recent experimental work on nanofiber mat cathodes for hydrogen/air fuel cells will be described. Procedures for fabricating high particle-loaded nanofibers containing various platinum, platinum-alloy, and platinum-group-metal free (PGM-free) catalysts will be presented. The effects of catalyst type and loading and MEA design on power output and durability in a hydrogen/air proton exchange membrane fuel cell at high and low relative humidity feed gas conditions will be discussed.

Acknowledgments

This work was funded by the Fuel Cell Consortium for Performance and Durability, DOE-EERE FC-PAD Project DE-EE0007653.

References

[1] Zhang and P. N. Pintauro, *ChemSusChem*, **4**, 1753-1757 (2011)

[2] Brodt, R. Wycisk, and P. N. Pintauro, *J. Electrochem. Soc.*, **160**, F744-F749 (2013).

[3] Brodt, T. Han, N. Dale, E. Niangar, R. Wycisk, and P. Pintauro, *J. Electrochem. Soc.*, **162**, F84-F91 (2015).

[4] Brodt, R. Wycisk, N. Dale, and P. Pintauro, *J. Electrochem. Soc.*, **163**, F401-F410 (2016).

[4] J. Slack, C. Gumeci, N. Dale, J. Parrondo, N. Macauley, R. Mukundan, D. Cullen, B. Sneed, K. More, P.N. Pintauro, *J. Electrochem. Soc.*, **166**, F3202-F3209 (2019).

354d The Impact of Inorganic Electrocatalyst Surface Ligands on Humidification Requirements in Operating Proton Exchange Membrane Fuel Cells

Ian Moorhead, John Williams, Anastasios Angelopoulos

Despite significant advances in the development of electrocatalysts with unprecedented Pt mass-specific electrocatalytic activity, the high cost of this precious metal remains an obstacle to proton exchange membrane fuel cell (PEMFC) fuel cell commercialization. A primary reason for this is that high-power applications drive thermodynamic irreversibilities beyond the catalytically-driven reaction kinetics associated with fundamental benchtop screening studies. Key among these irreversibilities is the transport of product liquid water that can obstruct gaseous reactant access to the electrocatalyst yet is at the same essential to membrane proton transport. Unlike industrial PEMFC, fundamental benchtop studies employ reagents dissolved in liquid electrolytes.

In this work, we address the question of how the interfacial chemistry of electrocatalyst nanoparticles used in electrodes can influence water management in industrial PEMFC. Our work employs both carbon supported and unsupported (carbon free) platinum-bismuth alloy nanoparticles (PtBi NP) since these have been shown to exhibit superior oxygen reduction reaction (ORR) activity and durability in fundamental benchtop investigations. For example, after 10000 potential cycles in accelerated durability testing, the PtBi NP demonstrate no degradation in mass activity, half-wave potential, Tafel slope and electrochemical surface area (ECSA). We employ a unique inorganic redox synthesis scheme to introduce hydrophilic stannous chloride ligands onto the surface of these nanoparticles with a well-characterized structure and charge density. Alloy nanoparticle surface hydrophilicity is shown to be key to optimizing the humidification and performance of an operating low temperature hydrogen-air PEMFC while maintaining durability at 25 cm2 scale.

354e Impact of Droplet/Bubble Growth Dynamics on Electrode Limiting Current

John Petrovick, Arthur Dizon, Clayton Radke, Adam Weber

Electrochemical devices often involve multiple phases, whether it is liquid-water droplets in hydrogen fuel cells or hydrogen and oxygen bubbles in water electrolyzers. The formation of these phases is directly coupled to the reaction sites within the catalyst layers, thus a complex multi-phase flow problem develops in these devices. In fuel cells, the production of water forms droplets that can retard the flow of reactant gas and make the catalyst inactive, dramatically reducing device performance. The problem is less egregious in electrolyzers, where the production of gas forms bubbles on the electrode surfaces, which eventually grow large enough to detach and exit the catalyst layer and may impact reactant transport to the underlying catalyst surface. In this work, we present a model for the transport of reactant (oxygen or water) through a growing droplet (bubble) of liquid (gas) and examine how the reaction rate or current density at the surface is impacted. The droplet grows primarily through water production at the electrode surface, while the bubble grows via two primary mechanisms, oxygen production at the electrode surface and supersaturation of oxygen in the water surrounding the bubble. Key components also include a force balance for size, transport within and outside the domain including the induced flux at the catalyst interface, and vapor-liquid equilibrium at the surface. The diffusion equation in cylindrical coordinates is used to capture the transport processes. This model provides significant insight into remaining transport inefficiencies in fuel cells and electrolyzers and how they can be mitigated.

354f A DFT+U Study of the Electrochemical Oxidation of H_2 and CO on $SrLaFeO_4$

Nicholas Szaro, Andreas Heyden

Solid oxide fuel cell (SOFC) anodes are commonly composed of cermet materials, typically Ni-YSZ. Single-phase electrodes such as the Sr-based Ruddlesden-Popper (RP) family serves as a baseline stoichiometry. Bulk doping strategies and nanoparticle exsolvation are used to tune activity. In this study, we elaborate on the mechanism of H_2 and CO electro-oxidation on $SrLaFeO_{4-}$ (SLF). We used *ab initio* methods to model H_2 and CO oxidation on two FeO_2-based surface models of SLF with surface Co and Ni.

All calculations were performed using the spin-polarized DFT+U method in VASP 5.4.4. We use PBE and the U-J values of 4.0 (Fe), 3.32 (Co), and 6.0 (Ni) eV. Slab models used a kinetic energy cutoff of 700 eV, a 3 × 3 × 1 Monkhorst-Pack mesh, a 15 Å vacuum layer, and the bottom two layers were fixed to bulk position values.

We tested all non-symmetrical conformers of SLF for the lowest relative energy on a model 2 × 2 × 1 supercell (Figure 1a). We developed two 1.5 × 1.5 × 1 non-identical (001) FeO_2-based slab models – FeO_2-LaO and FeO_2-SrO (Figure 1b). The oxidation mechanism of H_2 begins with a dissociative adsorption step. Subsequent steps include formation of surface H_2O, a surface vacancy, and a subsurface vacancy. We computed the bulk vacancy migration of SLF and assumed fast cathode kinetics to close the catalytic cycle. At $V_{cell} = 0.7$ V, we predict that the highest rate occurs on the FeO_2-SrO surface. We find surface H_2O formation to be rate-determining. Co and Ni doping decrease the barrier of the H_2 dissociative adsorption and surface H_2O formation steps. The oxidation mechanism of CO begins with adsorption of the CO molecule where CO migrates to a neighboring Fe-O complex to form surface CO_2. We find surface CO_2 formation to be the rate-determining step.

FIGURE 1: a) 2×2×1 supercell model of I4mm $SrLaFeO_4$, and b) Slab models of (001) FeO_2-LaO and FeO_2-SrO

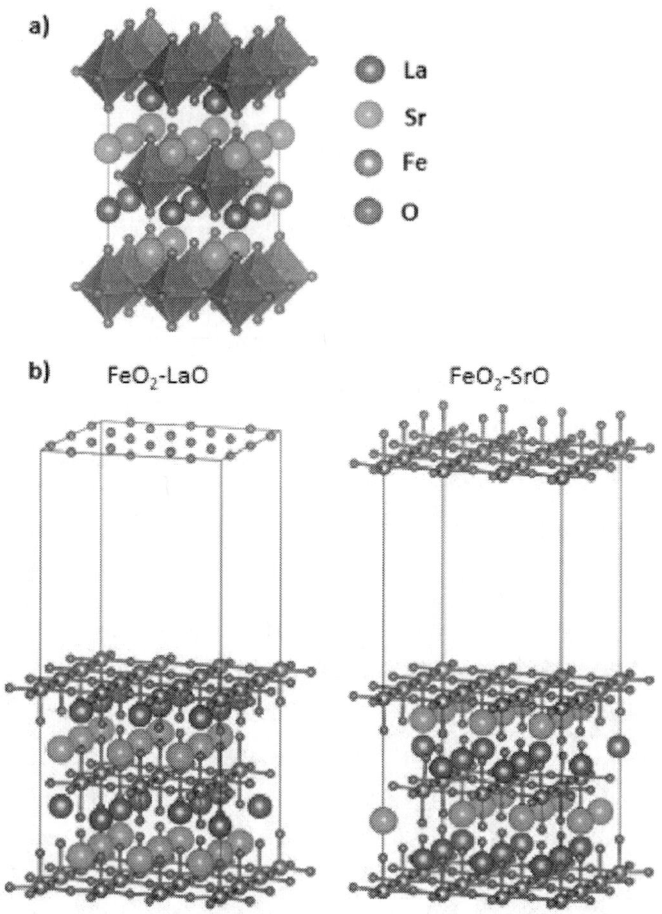

354g Durable and Highly Active PGM-Free Cathode Catalysts in Competition with Platinum for Pemfcs

Gang Wu

Atomically dispersed and nitrogen-coordinated single iron sites (FeN4) embedded in carbon (Fe-N-C) catalysts are the most promising platinum group metal (PGM)-free catalysts. However, they have yet to match their Pt counterparts for oxygen reduction reaction (ORR) activity and stability in proton exchange membrane fuel cells (PEMFCs). Here, we developed viable Fe-N-C catalysts, which, for the first time, demonstrated competitive activity to that of Pt/C catalysts and dramatically enhanced stability and durability under practical PEMFC operating conditions. The most active Fe-N-C catalyst achieved a record half-wave potential (E1/2 = 0.915 V vs. RHE at 0.6 mgcatcm-2) and an ORR mass activity of 10.5 mA mgcat at 0.9 V in (RDE) tests, exceeding a Pt/C baseline catalyst (60 μgPt cm-2) by 40 mV in acidic electrolytes. This compelling activity of the Fe-N-C catalyst in aqueous acids on rotating disk electrode (RDE) was successfully transferred to fuel cell membrane electrode assembly (MEAs), generating an initial current density of 44.2 mA cm-2 exceeding the U.S. DOE 2025 target (i.e., 44 mA cm-2) at 0.9 VIR-free under O2. Under practical hydrogen-air conditions, record 151 mA cm-2 at 0.8 V and peak power density of 601 mW cm-2 were achieved. Importantly, we discovered that depositing nitrogen-carbon species on the catalyst surface via chemical vapor deposition (CVD) dramatically enhanced catalyst stability, evidenced by performance durability after accelerated stress tests (30 000 square-wave voltage cycles under H2/air) and long-term steady-state life tests (> 300 hours at 0.67 V). Innovative identical location-scanning transmission electron microscopy (IL-STEM) experiments confirmed that the CVD process leads to deposition of nitrogen-doped carbon onto the catalyst surfaces. Along with theoretical modeling, a reconstruction of the carbon structure adjacent to FeN4 sites leads to increased robustness against demetallation and carbon oxidation. This work opens new avenues for developing earth-abundant iron-based catalysts with extraordinary activity and stability, thus competing with Pt and addressing the cost barrier of current PEMFCs.

577a Hydrogen Evolution Mitigation in Iron-Chromium Redox Flow Batteries Via Electrochemical Purification of the Electrolyte

Charles Wan, Kara Rodby, Mike Perry, Yet-Ming Chiang, Fikile R. Brushett

Redox flow batteries (RFBs) are a promising electrochemical storage technology due to their decoupling of energy and power, long service life, and scalability.[1] However, despite moderate commercial success, the widespread deployment of the state-of-the-art system, vanadium RFBs, is stymied by high capital costs. In particular, the high and volatile price of vanadium remains a significant obstacle to adoption, as the cost of the electrolyte alone (ca. 125 $/kWh, with significant fluctuations based on raw materials costs)[2] is close to the United States Department of Energy targets for the total installed system costs (generally cited between 100 and 150 $/kWh).[3,4] This has spurred research into RFB chemistries that utilize abundant and inexpensively-sourced active materials. Arguably the first contemporary RFB system, the iron-chromium (Fe-Cr) chemistry was originally advanced as an energy storage system for deep space missions in the 1970s. This chemistry has recently experienced renewed interest for grid storage as it utilizes relatively low-cost, high-abundance active species, and can remediate crossover losses by utilizing a mixed electrolyte configuration (i.e., the "spectator strategy"). However, among other challenges, a critical technical barrier inhibiting widespread Fe-Cr adoption is the high rate of the hydrogen evolution reaction (HER), which reduces system efficiency and causes severe capacity loss. Accordingly, research efforts in literature have primarily focused on catalyst development and electrolyte formulation to reduce HER rates.[5] In this presentation, we develop a protocol for electrochemical purification of the Fe-Cr negative electrolyte by reducing out metal impurities and explore the effect of the protocol on the performance of an Fe-Cr RFB single cell. We show that the purification process results in a notably reduced fade rate in durational galvanostatic cycling of an Fe-Cr RFB cell, and that the effectiveness of the protocol is dependent on the relative amount of electrolyte volume purified to the electrode area. Further, we extract performance metrics (i.e., coulombic, voltaic, and energy efficiencies along with capacity decay rate) of cycled Fe-Cr RFBs reported in the peer-reviewed literature, illuminating a correlation between coulombic efficiency and capacity decay rate. Following this trend, the performance of our cell using purified electrolyte is comparable to the performance of other cells using expensive catalysts and additives, evincing a potential cost reduction pathway for Fe-Cr RFBs.

Acknowledgments

This work was supported as part of the Joint Center for Energy Storage Research, an Energy Innovation Hub funded by the U.S. Department of Energy, Office of Science, Basic Energy Sciences. CTW acknowledges additional funding from the National Science Foundation Graduate Research Fellowship Program under Grant No. 1122374. Any opinions, findings, and conclusions or recommendations expressed in this material are those of the author and do not necessarily reflect the views of the National Science Foundation.

References

1. K. J. Kim, M.-S. Park, Y.-J. Kim, J. H. Kim, S. X. Dou, and M. Skyllas-Kazacos, *J. Mater. Chem. A*, **3**, 16913–16933 (2015).

2. K. E. Rodby, T. J. Carney, Y. Ashraf Gandomi, J. L. Barton, R. M. Darling, and F. R. Brushett,

Journal of Power Sources, **460**, 227958 (2020).

3. US Department of Energy, *Grid Energy Storage*, (2013).

4. ARPA-E, *ARPA-E: The First Seven Years*, (2016).

5. C. Sun and H. Zhang, *ChemSusChem*, **15**, e202101798 (2022).

577b Theory of Layered-Oxide Cathode Degradation in Li-Ion Batteries By Oxidation-Induced Cation Disorder

Debbie Zhuang, Martin Z. Bazant

Disorder-driven degradation phenomena, such as structural phase transformations and surface reconstructions, can significantly reduce the lifetime of Li-ion batteries, especially those with nickel-rich layered-oxide cathodes. Here, we develop a general free energy model for layered-oxide ion-intercalation materials as a function of the degree of disorder, which represents the density of defects in the host crystal. The model accounts for defect core energies, long-range dipolar electrostatic forces, and configurational entropy of the solid solution. In the case of nickel-rich oxides, we hypothesize that disordered nickel is driven into the bulk by electrostatic forces as oxidation reactions at the solid-electrolyte interface remove oxygen ions at high potentials (>4.3V vs. Li/Li+). The model is used in battery cycling simulations to describe the extent of cathode degradation when using different voltage cutoffs when cycling, in agreement with experimental observations that lower-voltage cycling can substantially reduce cathode degradation. The theory provides a framework to guide the development of cathode compositions, coatings and electrolytes to enhance rate capability and enhance battery lifetime.

577c Low Cobalt NCM811 with Cobalt Gradient for Improved Structural Stability and Electrochemical Performance

Arjun Patel, Sourav Mallick, Mingyao Mou, Jethrine Mugumya, Michael L. Rasche, Herman Lopez, M. Parans Paranthaman, Mo Jiang, Ram Gupta

Synthesis of low cobalt content $LiNi_{1-x-y}Co_xMn_yO_2$(NCM)-based cathode materials for high performance Lithium-ion batteries (LIBs) is need of this hour.[1] Among them, Nickle-rich layered oxide, NCM811 is regarded as the future of efficient cathode material for LIBs owing to its higher specific capacity and low cost. However, the commercialization of Ni-rich cathode material is hampered due to capacity fading from Li-Ni cation mixing and safety concerns related to the uncontrolled rise in temperature.[2] Cobalt (III) in NCM811 reduces the surface energy of the material and cationic disorder and a controlled distribution of cobalt from core to surface of the particle can improve the structural robustness of the NCM811, which further ensures the improved cycling performance of the LIBs. [3]

In this work, we synthesized layered NCM811 with cobalt gradient from core to the surface. A series of precursors with varying concentration of cobalt were prepared using coprecipitation method. The precursors were mixed with $LiOH{\cdot}H_2O$ (the molar ratio of Li:Ni–Co–Mn = 1.05:1) and calcinated at optimized temperature for 10 hrs. to achieve the final active material (NCM811). The electrochemical performance of synthesized full concentration gradient material with cobalt gradient was further compared with the homogeneous NCM811 in terms of specific capacity, energy density and cycling performance.

References

1. Schmuch, R., Wagner, R., Hörpel, G. *et al.*Performance and cost of materials for lithium-based rechargeable automotive batteries. *Nat Energy* **3,** 2018, Pages 267–278.
2. Longshan Li, Dingming Wang, Gaojie Xu, Qian Zhou, Jun Ma, Jianjun Zhang, Aobing Du, Zili Cui, Xinhong Zhou, Guanglei Cui, Recent progress on electrolyte functional additives for protection of nickel-rich layered oxide cathode materials, Journal of Energy Chemistry, Volume 65, 2022, Pages 280-292,
3. Zhenlei Huang, Jian Gao, Xiangming He, Jianjun Li, Changyin Jiang, Well-ordered spherical $LiNi_xCo_{(1-2x)}Mn_xO_2$ cathode materials synthesized from cobolt concentration-gradient precursors, Journal of Power Sources, Volume 202, 2012, Pages 284-290

577d Batch and Slug-Flow Synthesis of NMC111 Precursor Micro Crystals for Battery Application: A Comparative Study

Jethrine Mugumya, Mingyao Mou, Arjun Patel, Sourav Mallick, Michael L. Rasche, Ram Gupta, Mo Jiang

Increased interest in lithium ion batteries (LIBs) as a clean energy source is because of the growing demand for renewable energy alternatives to fossil fuels. Among the various electrochemical energy storage devices, lithium ion batteries are advantageous in terms of high energy and power density, rate capability, long cycle life and have a wide range of applications, including portable electronic devices, electric and hybrid electric vehicles.[1] Layered metal oxides such as NMC-111 ($LiNi_{1/3}Mn_{1/3}Co_{1/3}O_2$) are the popular choice for cathode materials because of their stability, high capacity and relatively low cost.[2] The electrochemical performance of NMC-111 largely depends on the composition, morphology and size distribution of the particles, which can be determined by proper modulation of the synthesis methods. Co-precipitation of precursor for NMC materials is commonly used because of scalability, controllability for the particle composition and morphology.[3] Here we compare co-precipitation synthesis of NMC-111 precursor materials from different reactors, traditional batch and new slug flow reactors with potential use[4,5]. The effect of reaction parameters on product crystal morphology and nucleation and growth kinetics are quantified.

References.

(1) Hua, W.; Liu, W.; Chen, M.; Indris, S.; Zheng, Z.; Guo, X.; Bruns, M.; Wu, T. H.; Chen, Y.; Zhong, B.; Chou, S.; Kang, Y. M.; Ehrenberg, H. Unravelling the Growth Mechanism of Hierarchically Structured $Ni_{1/3}Co_{1/3}Mn_{1/3}(OH)_2$ and Their Application as Precursors for High-Power Cathode Materials. *Electrochim. Acta* **2017**, *232*, 123–131. https://doi.org/10.1016/j.electacta.2017.02.105.

(2) Manthiram, A. A Reflection on Lithium-Ion Battery Cathode Chemistry. *Nature Communications*. 2020. https://doi.org/10.1038/s41467-020-15355-0.

(3) Dong, H.; Koenig, G. M. A Review on Synthesis and Engineering of Crystal Precursors Produced: Via Coprecipitation for Multicomponent Lithium-Ion Battery Cathode Materials. *CrystEngComm* **2020**, *22* (9), 1514–1530. https://doi.org/10.1039/c9ce00679f.

(4) Jiang, M.; Zhu, Z.; Jimenez, E.; Papageorgiou, C. D.; Waetzig, J.; Hardy, A.; Langston, M.; Braatz, R. D. Continuous-Flow Tubular Crystallization in Slugs Spontaneously Induced by Hydrodynamics. **2014**. https://doi.org/10.1021/cg401715e.

(5) Guillemet-Fritsch, S.; Aoun-Habbache, M.; Sarrias, J.; Rousset, A.; Jongen, N.; Donnet, M.; Bowen, P.; Lemaître, J. High-Quality Nickel Manganese Oxalate Powders Synthesized in a New Segmented Flow Tubular Reactor. *Solid State Ionics* **2004**, *171* (1–2), 135–140. https://doi.org/10.1016/S0167-2738(03)00282-0.

577e High Power Symmetric Zinc Iodine Redox Flow Batteries with Functionalized Carbon Electrodes

Abena Williams, Robert Emmett, William Marcengill, Mark E. Roberts

Redox flow batteries (RFBs) have emerged as promising candidates for low cost energy storage to support renewable energy generation. Vanadium redox flow batteries (VRFB) are the most widely commercialized RFB but are restricted by expense due to the high cost and limited availability of vanadium and corrosion due to the highly acidic electrolyte. The zinc iodine (ZnI_2) redox flow battery has great potential as an alternative to VRFB because it consists of low cost, environmentally friendly and widely available materials. Additionally, it is a symmetric battery which has a highly electroactive electrolyte on both sides of the battery which minimizes electrolyte crossover. Previously we showed that carbon electrodes modified with redox-active iron nanoparticles yield increased rates of charge transfer, improved voltage efficiencies at elevated current densities and higher power densities in coin cells using iron based redox electrolytes. Here we show that using iron modified carbon electrodes in a zinc iodine redox flow battery increases the charge transfer kinetics and power capabilities. The economic value and enhanced electrochemical performance can rival the vanadium chemistry and makes the ZnI_2 RFB a viable option for large scale electrochemical energy storage applications.
In this work, we studied how carbon electrodes functionalized with redox active iron particles affect the performance of zinc iodine redox flow batteries. Small scale ZnI_2 RFBs were assembled with pure carbon nanotube electrodes and then with carbon nanotubes containing iron particles electrodes. The electrodes were conditioned and then tested over a range of currents, flowrates, and electrolyte concentrations. Cyclic voltammetry and electrochemical impedance spectroscopy were also conducted to evaluate the performance of the redox electrolyte paired with the electrodes. Our ZnI_2 RFB with carbon electrodes containing redox-active iron particles exhibited lower charge transfer resistances, higher discharge voltages and were able to sustain currents of up to 700mA, which is a 28% increase relative to the system with pure carbon electrodes. Polarization analysis showed that iron electrodes demonstrated less effect of electrolyte flowrate on electrochemical performance and an approximately 50% increase (89mW/cm^2 to 44mW/cm^2) in power density at 110mA/cm^2 relative to pure carbon electrodes. The results indicate that iron modified electrodes improve the kinetics at the electrode-electrolyte interface and electrolyte utilization and thus can be employed in other RFB chemistries to reduce material costs and improve efficiency. The adaptation of functionalized carbon electrodes in low cost RFBs like the zinc iodine provides cost effective electrochemical energy storage for expansion of intermittent energy sources such as wind and solar to off grid applications.

577f Electrochemical-Mechanical Microstructure-Scale Battery Modeling for Improved Performance and Degradation Predictions during Fast Charging.

Francois Usseglio Viretta, Hari Sitaraman, Michael Brazell, Jeffery Allen, Marc Day, Kandler Smith

Battery performance is strongly correlated with electrode microstructure. Macro-homogeneous electrochemical models typically use porous electrode theory and thus abstract microstructural heterogeneity of composite battery electrodes using effective macroscopic properties [1]. Such models only require a small number of degrees of freedom, however they are intrinsically limited by their macroscale approach and related assumptions. In particular, in-plane heterogeneity is neglected, and particles are assumed spherical which, respectively, hinder degradation predictions and relevance for materials with more complex particle morphologies such as graphite.

Microstructure-scale models remedy these limitations by directly solving the governing equations on the electrode's constituent geometry [2]. Electrode materials for lithium-ion batteries have complex microstructure geometries that require millions of degrees of freedom to solve the system at the microstructure scale and be representative. In this work [3], a microstructure-scale coupled electrochemical-mechanical model is developed. The solver uses a level-set method for modeling electrode-electrolyte interfaces and surface reactions. It is built on a block-structured adaptive-mesh-refinement framework using open-source library, AMReX [4], and can currently utilize emerging heterogenous (CPU+GPU) computing architectures. This approach simplifies mesh generation for complex microstructure geometries and leverages fast and scalable multi-grid based linear solvers on Cartesian grids. Simulations of fast charging for automotive application, with electrode geometries obtained through Computed-Tomography imaging are presented (cf. figure below). Concentration gradients, in-plane heterogeneities, cell-voltage and potential for lithium plating are discussed and compared with a baseline macroscale model. In addition, stress distribution is analyzed within the electrode volume. The model demonstrates excellent mass conversation and high numerical scaling efficiency on high-performance computers (HPC), both for CPU and GPU architectures.

[1] F. Usseglio-Viretta et. al., *Journal of The Electrochemical Society*,165(14) A3403-A3426 (2018), https://doi.org/10.1149/2.0731814jes

[2] J. Allen et al., *Journal of Scientific Computing*, 86:42 (2021), https://doi.org/10.1007/s10915-021-01410-5

[3] Article in preparation.

[4] https://amrex-codes.github.io/

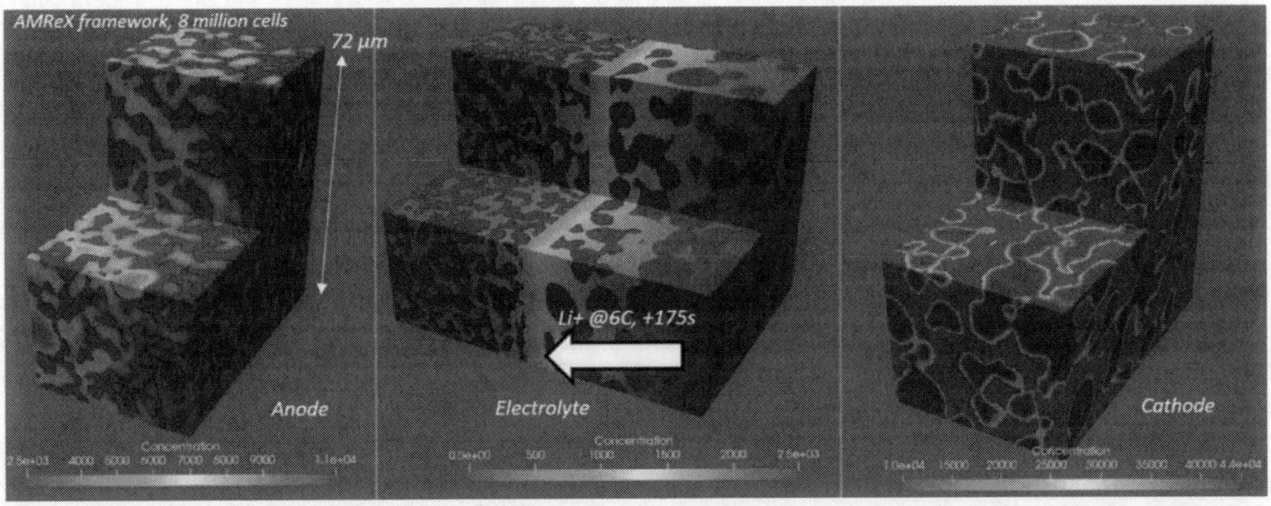

Concentration profile calculated within a graphite/NMC cell during fast charging. Geometry from X-ray Computed tomography.

577g A Molecular Simulation Study of the Solvation and Selectivity of Glycerol-Based Solvents for Different Cationic Species.

Gabriel Barbosa, Jason Bara, C. Heath Turner

Over the past few years, people have extensively studied the tendency of ether-based acyclic solvents to confine positively charged atoms, forming structures called podands. Due to the stability of these structures, podands have been identified as promising candidates for several different electrochemical applications. In this work, we used molecular dynamics simulations to investigate the solvation and selectivity of Na^+, K^+, and Li^+ by different glycerol-derived solvents. Additionally, hybrid Monte Carlo sampling is used to accelerate the configurational equilibration, in order to provide additional verification of the molecular dynamics equilibrium predictions. By analyzing the oxygen-cation site-site radial distribution functions, strong coordination between these sites is found. In addition, further analysis of the structural properties suggest that the solvent-cation association is stable over a very wide temperature range. A strong correlation is observed between the more electronegative oxygen sites and preferential interaction with the cations, which suggests that the electrostatic oxygen-ion interactions dominate the coordination between these species. Significantly more intense interactions are observed between the solvents and Li^+, versus Na^+ or K^+. These molecular-level thermodynamic insights provide fundamental support for further exploring the performance of glycerol-derived podand complexes in Li^+ capture and/or electrolyte formulations.

AUTHOR INDEX